Samuel da Silva Moraes
Mariana Pessôa Coelho
Elisabeth Regina Alves Cavalcanti Silva

The quality of Brazilian highways

Samuel da Silva Moraes
Mariana Pessôa Coelho
Elisabeth Regina Alves Cavalcanti Silva

The quality of Brazilian highways

A comparative study with the roads of the BRIC countries (Brazil, Russia, India and China)

ScienciaScripts

Imprint

Cover image: www.ingimage.com

This book is a translation from the original published under ISBN 978-613-9-68840-1.

Publisher:
Sciencia Scripts
is a trademark of
Dodo Books Indian Ocean Ltd. and OmniScriptum S.R.L publishing group

120 High Road, East Finchley, London, N2 9ED, United Kingdom
Str. Armeneasca 28/1, office 1, Chisinau MD-2012, Republic of Moldova, Europe
Printed at: see last page
ISBN: 978-620-8-23139-2

SUMMARY

ACKNOWLEDGMENTS

First of all, we thank God for the opportunity to be in this world and to live with such special people.

Especially our family members who encouraged us to keep going even though we knew it was difficult, but not impossible.

To our friends from university who made us great with their knowledge and their friendships.

To the Joaquim Nabuco College, for the opportunity to broaden our knowledge, and especially to our teachers, who helped us every step of the way.

To our advisor, Professor Adeilde Santana, who guided us in the right direction and with that we were able to achieve our dreams and goals.

To Master Mariana Pessôa Coelho, who was fundamental to the development of this final work. Our sincere thanks to you.

SUMMARY

Brazil, a country with a large territorial area, has the main route for transporting its production by road, which in most parts of the country faces numerous difficulties due to neglect by the authorities, poor infrastructure and a wide range of other problems. Based on a comparative study of countries with similar economic and geographic situations to Brazil, the aim of this work is to address some points of the utmost importance in the search for alternatives to the problems encountered on Brazil's federal highways. In view of the proposed work, studies and research were carried out into the current state of the Brazilian road structure, gaining a better understanding of the types of road, demonstrating the advantages and disadvantages of the road modal and pointing out the disparity between roads granted to the private sector and stretches managed by the Brazilian government, as well as providing a quantitative and qualitative comparison of the road transport of countries such as Russia, India and China, members of the BRICS, with South Africa being excluded from this study as it does not have similar geographical characteristics to the other members. This was followed by a survey of the main products traded between this economic group and the recent growth in exports and imports within the BRIC, pointing to an evolution in trade between these countries.

Key words: Infrastructure. Road modal. BRIC.

INTRODUCTION

Brazil has a continental territory, and this alone makes it a huge challenge for transport logistics in the national territory, where the globalized market seeks to reduce costs and delivery times to meet the demands of increasingly discerning customers in their choices.

In the early 2000s, economist Jim O'Neal used the term BRIC for the first time to list the countries with the greatest potential for world economic growth. In his view, Brazil, Russia, India and China formed a bloc of countries with a stabilized and constantly growing economy. Today, this group of countries is called BRICS, with the addition of South Africa in 2011, but it was excluded from this study because it does not have the same geographical characteristics and territorial dimensions as the other components.

In countries with large geographic dimensions, such as the BRIC countries, there are a large number of interconnected highways to take development to all the limits of their territory, especially in countries that do not use the multimodal concept as a principle for logistics activities, relying on a single modal for the flow of products in general, increasing costs and the risk of constant losses in their transportation.

This paper will present a survey of the main Brazilian highways, with a qualitative and quantitative approach to the current road situation in Brazil and a better understanding of this mode of transportation in the other BRIC countries, as well as seeking to understand the reality found on Brazilian roads.

1. OBJECTIVE

Carry out a survey of the main Brazilian highways, in terms of transportation of crops and products sold to the domestic and foreign markets, comparing them with the infrastructure of the emerging countries that make up the BRIC and the current state of Brazilian transportation.

1.1 Specific objectives:

- To gather information on the length (km) of BRIC roads.
- Comparing the conditions of use of toll and non-tariff roads in Brazil.
- Identify types of Brazilian road construction and compare them with those of the BRIC countries.
- Identify the most transported products in BRIC and their consequences for roads.
- Investigating the causes of wear and tear on BRIC roads

2. METHODOLOGY

By means of a bibliographical survey, we obtained information about the economy and social development involving the BRIC, about the common characteristics within these countries.

In this way, quantitative and qualitative information on Brazilian highways was also obtained, together with a search on data made available on the internet by the federal government, and various information was found on the Brazilian road modal regarding the main products transported, damage caused to the Brazilian road network and demonstrating the disparity in the administration of public highways in relation to stretches concessioned to the private sector.

And through exploratory research, facts were obtained about the other BRIC countries, such as the quality of some roads and peculiarities about the road modal in these countries, in order to make a superficial comparison with the daily facts found in Brazil.

3. THEORETICAL FRAMEWORK

3.1 BRIC members and the origin of the term

The British economist Jim O'Neill coined the term BRIC in 2001, referring to the emerging countries of Brazil, Russia, India and China, which had similar characteristics in terms of economic growth, gross domestic product (GDP), reduction of social disparities and investments in foreign companies in various sectors of the economy (CASELLA, 2010). According to the author, the BRIC countries have managed to look after themselves and create new strategies for entering other international markets.

According to Casella (2010), the BRICS member countries have points in common, as described below:

- Recently stabilized economy;
- Stable political situation;
- A large and qualified workforce;
- Growing production and export levels;
- Good reserves of mineral resources;
- Investments in infrastructure sectors (roads, railroads, ports, airports, hydroelectric plants, etc.);
- GDP (Gross Domestic Product) on the rise;
- social indices in the process of improvement;

- Reduction, albeit slow, of social inequalities;

- Rapid access of the population to communication systems such as cell phones and the Internet (digital inclusion);

- Capital markets (stock exchanges) receiving large foreign investments;

- Investments by foreign companies in the various sectors of the economy.

Another striking fact for understanding the characteristics of these countries is the analysis of the socio-economic data shown in Table 1 below.

Table 1 - Socio-Economic Data BRICS Member Countries

Socio-economic data	Brazil	Russia	india	China	Africa South
HDI (Human Development Index)	0,744	0,778	0,586	0,719	0,658
Life expectancy at birth	73.9 years	68 years old	66.4 years	75.3 years	56.9 years
Average years of schooling	7.2 years	11.7 years	4.4 years	7.5 years	9.9 years

Source: UNDP (2013). Prepared by the authors

For the United Nations Development Program (UNDP), the Human Development Index (HDI) is an indicator that, through three basic conditions, synthesizes the complex dimensions of human development.

Longevity indicators use life expectancy at birth. As an indicator of educational attainment, we use the literacy rate and the combined enrollment rate in primary, secondary and higher education. Finally, per capita income is calculated to obtain an indicator of access to resources. The final HDI result will be between 0 and 1. It is analyzed as follows:

- Results between 0 and 0.5 Considered a low level of development
- Results between 0.5 and 0.8 Considered an average level of development
- Results between 0.8 and 1 Considered a high level of development

Brazil, with a population of 207.8 million, is currently facing a crisis as a result of political problems. It has a strong and solid economy, and is a major producer and exporter of various types of goods, especially agriculture, mineral commodities and manufactured goods. Agriculture, industry and services are well developed and are expanding.

It is the 9th largest economy in the world (2016) with a GDP of US$1.774 trillion according to the World Bank/2015, and has an open economy, being directly involved in the globalization process. It has an investment rate of 17.8% of GDP (2nd quarter 2015), and inflation stands at 10.67% according to the IPCA 2015 and an alarming increase in the unemployment rate, which stands at 12% of the economically active population.

Its main agricultural products are coffee, oranges, sugar cane (sugar and alcohol), soybeans, tobacco, corn and mate. All of which are extremely dependent on roads. With a land area of 8.51 million km,^{2} offers great agricultural plantations, but the transportation of its harvests depends basically on a single mode of transport that is, for the most part, deficient.

Russia, with a population of more than 144 million, is currently ranked eleventh in the world in terms of GDP, with a total of US$1.326 trillion (World Bank/2015), 20% of which goes on investment.

In the most recent survey in 2015, Russia had an unemployment rate below 6%. Its main economic sectors are industry, mining (oil and natural gases) and agriculture. It has the world's largest territory with more than 17 million km2, which poses major challenges for logistics processes in such a huge space, especially for internal transportation logistics.

The export of crude oil and oil derivatives, along with metals, natural gas and chemical products, are considered key to the Russian economy, resulting in a trade surplus of approximately US$140.5 billion.

India, like all the others, is an emerging country with around 1.2 billion inhabitants. In recent decades, its GDP has grown by 7% a year and it is now the world's seventh largest economy, with a GDP of US$2.073 trillion, according to the World Bank/2015.073 trillion, with a diversified economy, it has strong results in the agricultural, industrial and technological areas, the latter two being where the Indian government has invested the most in recent decades, and which today show fantastic results for the economy.

Among the BRIC countries, it has the smallest territorial area with 3.29 million km^2 , less than half of what Brazil has, for example, but even with this relatively small territory, it has more than four times as many paved roads as Brazil, according to the CIA (2016), thanks to the investments made in infrastructure in recent decades, which has boosted the Indian economy to the place it occupies today among the world's largest economies.

Among the members, the one that stands out the most is China, with a population of over 1.3 billion inhabitants, a GDP much higher than the others, according to the World Bank / 2015, approximately US$ 10. 982 trillion, and with a high growth rate in recent years, even with a small drop in these figures, it has achieved a 6.9% increase in its Gross Domestic Product.982 trillion, and with high growth in recent years, even with a small drop in these figures, it has achieved a 6.9% increase in its Gross Domestic Product, consolidating itself as the second largest economy in the world (behind only the United States), figures that make China the holder of 15% of the world economy.

Updated data from the World Bank shows that China's GDP figures are higher than those of the other BRIC members, as can be seen in Graph 1. Together, the four countries have a Gross Domestic Product of approximately US$16.155 trillion, with China having the greatest influence on this figure, as shown in Graph 2.

Graph 1: GDP volume in BRIC

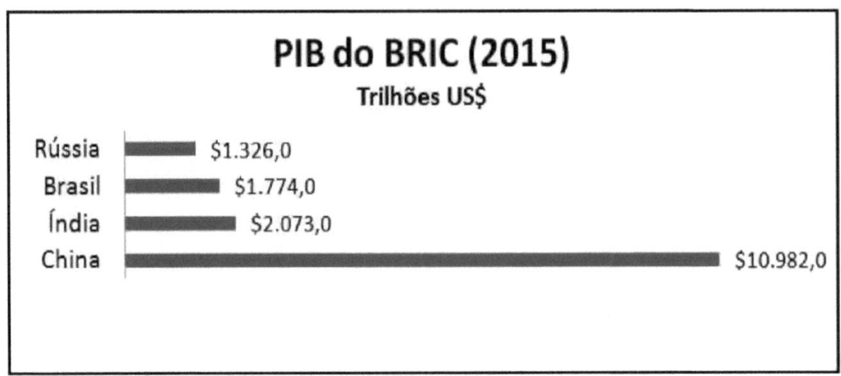

Source: World Bank/2015. Prepared by the authors.

Graph 2: BRIC's influence on GDP.

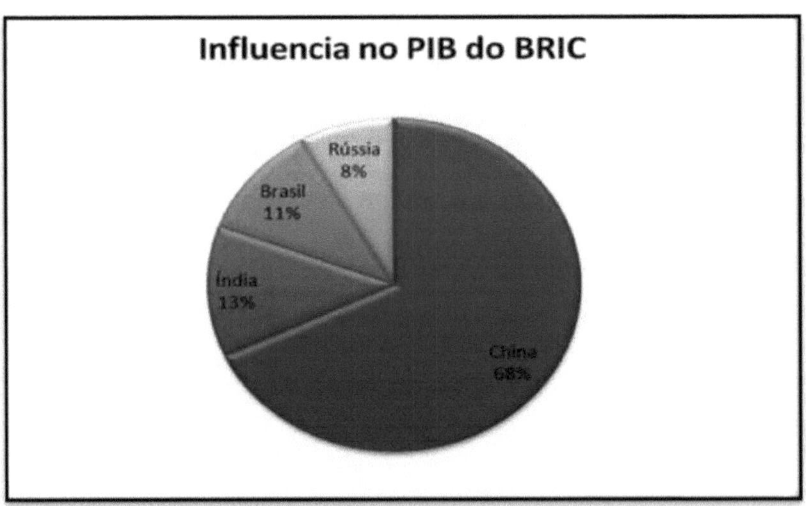

Source: World Bank/2015. Prepared by the authors.

With the opening up of its economy, multinational companies have set up their headquarters in China in search of abundant, low-cost labor. The world's largest food producer, investing in mechanized agriculture and generating excellent results, it is the world's largest producer of rice and corn.

With this accelerated growth in their economy, the Chinese are investing in infrastructure, with the expansion of roads, railroads and airports.

According to Santos (2010), of the BRIC countries, India and Brazil are the least dependent on the world market, because they are more dependent on domestic market demand. Russia, on the other hand, is very dependent on commodity prices, mainly oil and natural gas, so with the slowdown and decrease in the prices of natural resources, it tends to suffer the consequences of the crisis.

China, despite being more dependent on external demand than the other BRIC members, has adopted a strong program to stimulate domestic demand and has advanced into some non-traditional markets with its products.

3.2 Roads: characteristics, types and classification of use.

In recent decades, the process of urbanization of Brazilian cities has occurred in a disorderly manner, causing immense problems in terms of the quality of life of the population and also in terms of structural issues in the country's urban areas.

In Brazil, the most widely used means of transporting cargo is by land, with road being the main mode among all the other modes. It is characterized by its simplicity of operation, and stands out for transport over short and medium distances, and for higher value-added cargo (VIEIRA 2001 *apud* FERREIRA, 2011), but of all the modes it has the lowest load capacity. This mode has a good cost-benefit ratio and also less cargo handling, as it is possible to offer a door-to-door service, i.e. the cargo is taken from one point to another directly (point of departure to its final destination).

In countries as large as the BRIC countries, there are a large number of interconnected roads to take development to all the limits of their territory. However, much of Brazil's road network is in a poor state of repair.

In Brazil, according to the DNIT (National Department of Transport Infrastructure), there are five types of federal highways with nomenclature defined by the PNA (National Aviation Plan), as a common characteristic, they are identified by the acronym BR and followed by three digits, the first digit being the indicator of the category of the highway according to the definitions established by the PNA, and they are broken down below according to the DNIT.

> Radial highways: These are highways that run from the federal capital towards the extremes of the country, their first digit is 0 (zero).

Figura 1. BR 010 - Rodovia Bernardo Sayâo or Rodovia Belém-Brasilia (linking Brasilia-DF to Belém-PA).

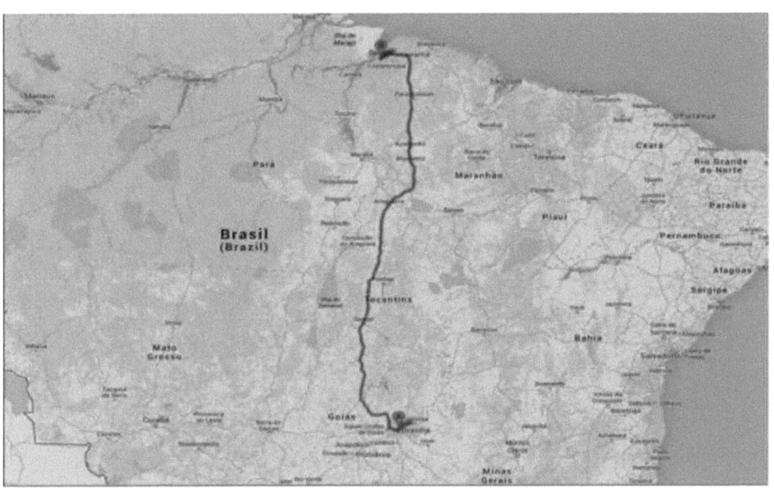

Source: Google Maps. Adapted by the authors

> Longitudinal Roads: These are the roads that cut across the country in a north-south direction, with the first digit being 1 (one).

Figura 2. BR 101 - Governador Mario Covas Highway (Touros - RN to Sâo José do Norte - RS).

Source: Google maps. Adapted by the authors

> Transverse Highways: These are the highways that cut across the country in an east-west direction, with the first digit being 2 (two).

Figura 3. BR-262 (linking Vitória-ES to Corumbâ-MS).

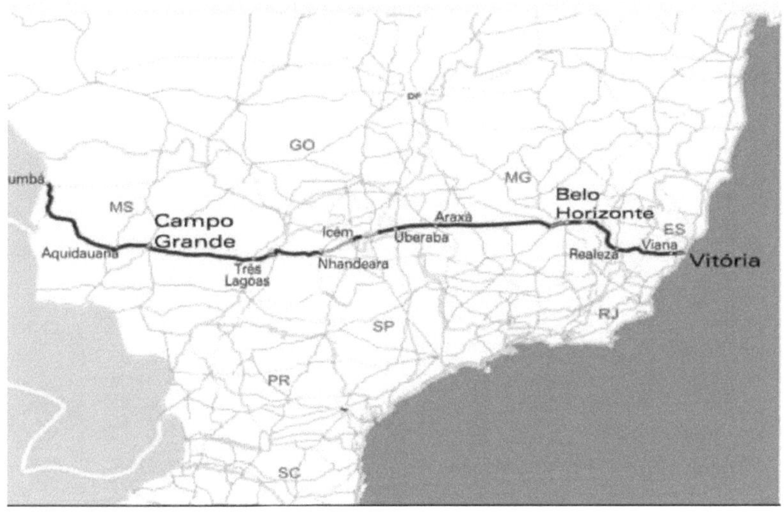

Source: Google Maps. Adapted by the authors

> Diagonal roads: These roads can be oriented in two ways: northwest-southeast or northeast-southwest, with the first digit being the number 3.

Figura 4. Diagonal roads oriented in a general NO-SE direction: BR-364 (linking Limeira-SP to Rodrigues Alves- AC).

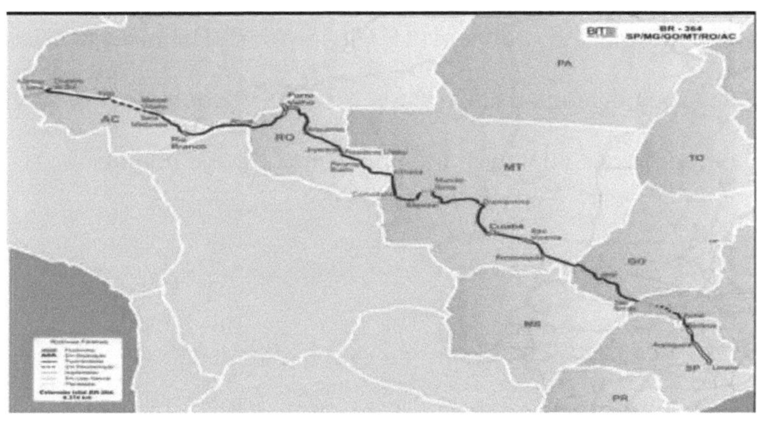

Source: Google Maps. Adapted by the authors

Figura 5. Diagonal highways oriented in a general NE-SO direction: BR-381 (linking Sâo Mateus-ES to Sâo Paulo-SP).

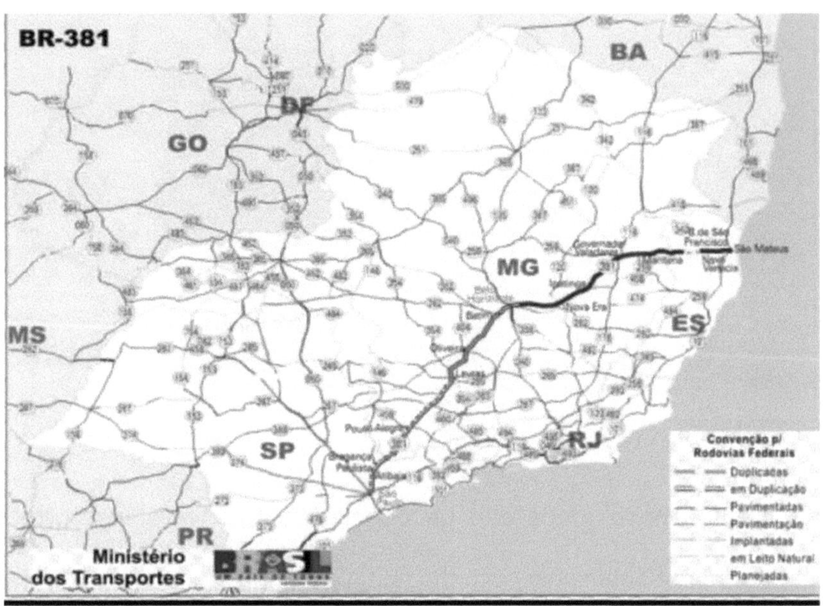

Source: Google Maps. Adapted by the authors.

> Link Roads: These roads do not have a specific direction, but are characterized by the fact that they connect at least one federal highway to cities or important points. Their mileage usually starts from the northernmost point of the road to the southernmost point, with the first digit being the number 4.

Figure 6. BR-470 (linking Navegantes-SC to Camaquâ- RS).

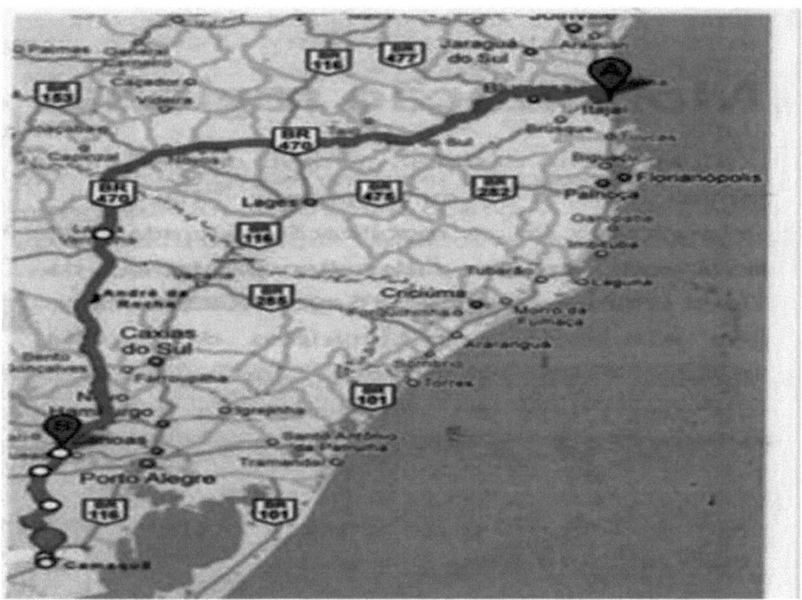

Source: Google Maps. Adapted by the authors

According to Ribeiro & Ferreira (2002) Integration between modes can occur between several modes: air-rail, rail-road, water-rail, water-road or even more than two modes.

The use of more than one mode adds advantages to each mode, characterized by the level of service and cost. Combined, they allow door-to-door delivery at a lower cost and a relatively shorter time, seeking a balance between price and service.

Figure 7. World Economic Forum Ranking - Transport Infrastructure Pillar (Selected Countries)

Countries	Transportation in General	Highways	Railroads	Ports	Ae roportos
Hong Kong SAR	1°	5°	3°	5°	3°
Singapore	2°	3°	8°	2°	1°
The Netherlands	3°	2°	7°	1°	4°
Finland	4°	12°	5°	4°	9°
Japan	5°	8°	1°	22°	25°
South Africa	26°	34°	42°	36°	14°
China	29°	42°	16°	50°	51°
India	47°	61°	29°	60°	71°
Russia	59°	123°	24°	75°	77°
Argentina	89°	108°	93°	81°	92°
Peru	92°	111°	94°	86°	82°
Colombia	103°	126°	106°	85°	74°
Brazil	115°	121°	98°	120°	95°
Venezuela	131°	128°	104°	125°	134°

Source: World Economic Forum (The Global Competitiveness Report 2015-2016)

3.3 Management of Brazilian roads x BRIC

The Brazilian government, through bodies such as the Department of Roads (DER), the National Land Transport Agency (ANTT) and the DNIT, are responsible for the road infrastructure within the national territory, and frequently make data, information and surveys on the condition of Brazilian roads available on their respective websites.

In Brazil, according to the CNT (National Transport Council), the road is the most important means of transporting cargo and is essential for the country's economy. Due to the lack of investment in road infrastructure, the government has created highway concessions so that, in exchange for tolls, private companies take on the responsibility of maintaining the quality of the existing infrastructure. The investments made by the union and state-owned companies in billions and by modal in 2014 are shown below, with highways showing the greatest investment (Figure 8):

Figure 8. Investments by the Union and State-owned Companies in billions and by modal

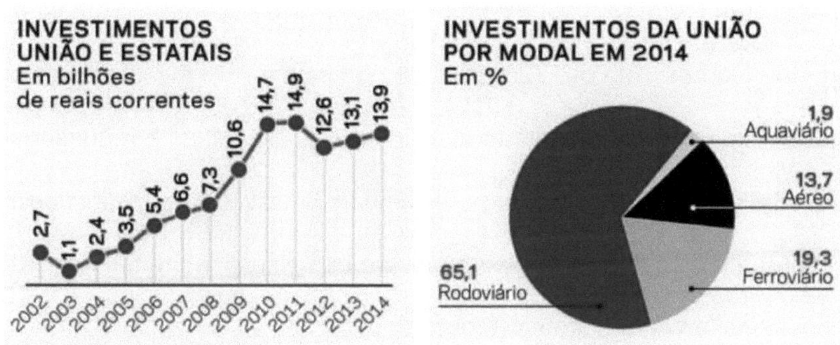

Source: CNT.

The Federal Highway Concession Program covers 11,191.1 kilometers of highways, broken down into concessions promoted by the Ministry of Transport and State Governments, through delegations based on Law No. 9,277/96, and by ANTT.

Art. 1 - The Union, through the Ministry of Transport, is hereby authorized to delegate, for a period of up to twenty-five years, extendable for a further twenty-five, to municipalities, states of the Federation or the Federal District, or to a consortium between them, the administration of highways and the operation of stretches of highways, or federal road works.

(...)Art. 3 The delegation shall be formalized by means of an agreement. (...) § 1 The agreement instrument shall include a clause providing for the possibility of applying the legislation of the Municipality, State or Federal District in the collection of tolls or port tariffs, or any other applicable form of collection, insofar as it does not contradict federal legislation.

According to Gomes et al. (2003), the objective of the concession encompasses expectations of increased road safety, as well as a reduction in total transportation costs (fuel savings, reduction in travel time, reduction in vehicle maintenance costs). As an example, we can visualize the tolls per KM of 20 concessions in various states (Figure 9).

Figure 9. Toll values per KM of 20 concessions in various states

Posição	Concessionária	Trecho	R$/KM
1	Rota dos coqueiros	**6,52 km** em Jaboatão dos Guararapes (PE) – urbano	0,874
2	Viario	**13 km** no Rio de Janeiro (Transolímpica) – urbano	0,558
3	Rota do Atlântico	**35,2 km** que leva ao Porto de Suape (PE)	0,415
4	Lamsa	**17,4 km** no Rio de Janeiro (Linha Amarela) – urbano	0,413
5	Concer	**180,4 km** ligando Rio de Janeiro a Minas Gerais	0,206
6	Ecovias	**176,7 km** - Sistema Anchieta-Imigrantes – SP	0,199
7	Via Lagos	**57 km** no Rio de Janeiro	0,198
8	Autoban	**319,8 km** – Sistema Anhanguera-Bandeirantes – SP	0,196
9	Centrovias	**218,16 km** na região centro-norte de SP	0,181
10	Bahia Norte	**121,4 km** de Salvador ao Litoral Norte – BA	0,181
11	**Econorte**	**343,7 km no Norte do Paraná**	**0,178**
15	SPVias	**505,7 km** de estradas incluindo Castelo Branco – SP	0,165
19	**Caminhos do Paraná**	**405,9km de Guarapuava a Araucária – PR**	**0,149**
25	Novadutra	**402 km** da Via Dutra, entre Rio e São Paulo	0,141
26	**Ecocataratas**	**458,9 km entre Guarapuava e Foz do Iguaçu**	**0,141**
29	**Viapar**	**550,5 km entre Arapongas e Cascavel – PR**	**0,127**
31	**Rodonorte**	**567,7 km entre Apucarana e Curitiba – PR**	**0,122**
34	**Ecovia**	**175 km entre Curitiba e Paranaguá**	**0,111**
47	Regis Bittencourt	**401,6 km** entre São Paulo e Curitiba	0,046
49	Fernão Dias	**562,1 km** entre Guarulhos (SP) e Contagem (MG)	0,033

Source: ABCR

Since Brazil cedes stretches of its highways to private companies for concessions, the latter, through tolls, raise revenue for the maintenance of these stretches, with a view to improving quality, comfort and safety (Figure 10), outperforming roads that don't have any kind of concession and depend exclusively on the government for maintenance (Figure 11).

Figure 10. Privatized stretch of Rod. Régis Bitencourt / BR116

Figure 11. Non-privatized stretch of the BR116 highway

With regard to highway concessions, China faces a problem with excessive tolls (Figure 12) and abusive charging. According to Epochtimes/2014, tolls on Chinese highways are among the highest in the world, charging much more per kilometer than is normal in developed countries. According to reports, "relevant departments" have announced that net national toll collection in 2010 amounted to 280 billion Yuan (c. US$45 billion).

Audits have revealed a number of irregularities and unnecessary expenses, such as hiring an excessive number of employees when the need was less than requested.

What is clear is that the Chinese highway system has established itself as yet another interest group in cahoots with the state authorities and the Chinese Communist Party to perpetuate the exploitation of the population and maximize the waste of resources.

Figure 12. Excessive number of tolls on Chinese highways.

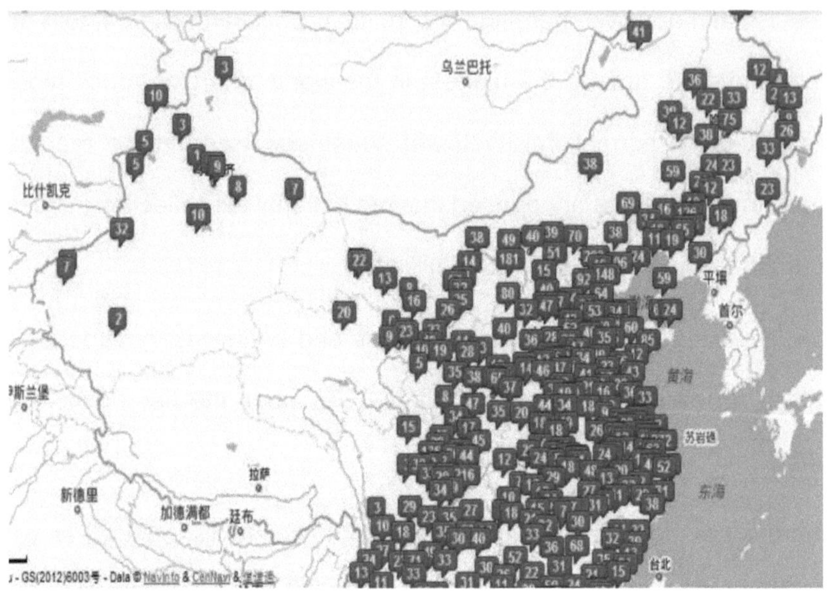

Source: Epochtimes/2014

In India, a serious problem to be pointed out in this work is the number of deaths that occur on Indian roads. The WHO/2014 points out that every five minutes in Indian traffic, one person dies.

According to Philip/2014, this significant number of deaths exposes the negative side of motoring in India: high levels of traffic jams in major centers, drivers without adequate training, lack of enforcement of traffic laws and poorly designed roads, coupled with the fact that the most fatal routes in the world are located in India, and even with imminent problems on its highways, in recent years there has been an evolution with regard to major highways in India, and today it is possible to contemplate exceptional works such as the Mumbai highway (Figure 13) and the Mumbai Pune Expressway (Figure 14).

Figure 13. Central highway in Mumbai

Source: Exame online magazine /2014.

Figure 14. Mumbai Pune Expressway

Source: Skyscrapercity/2013.

The extremely cold climate found in Russia poses a major problem for the movement of cargo on Russian roads, putting drivers in front of adversities such as the accumulation of snow on the roads and the loss of vehicle stability, often resulting in accidents on Russian roads.

Another factor that hinders Russian traffic is the excessive number of trucks on the roads, often causing congestion and constant danger to other drivers, as can be seen in figure 15 below.

Figure 15. Moscow Ring Road

Source: The Moscow Times/2013.

This problem was the reason why, according to The Moscow Times (2013), the Moscow government put into practice a plan to ban large trucks from passing through Moscow's ring road, in order to alleviate the problems of heavy traffic jams and damage to the roads, even though it goes against the logistical vision of Russian experts, who are fighting to overturn this plan because they warn of the damage to the economy that measures like these can cause.

3.4 Brazil's road transport and the flow of products.

The current flow of products, whether within the national territory or destined for export, uses road, rail and river transport, and even though Brazil is a country with continental dimensions, it invests little in logistics infrastructure and relies on a road network with the smallest in the BRIC, lagging far behind the other members of the group.

According to the CNT/2014, road transport was responsible for transporting more than 485.625 million TKU (tons transported per useful kilometer), around 61% of the country's cargo. And even though road transport is the main mode of transport in Brazil, some roads are in a poor state of repair, less than desirable for those who use this mode, significantly increasing vehicle maintenance costs and contributing to a considerable increase in the risk of accidents.

Studies show that road transportation over medium to long distances is costly, with all the costs involved in transporting the product, which are added to the product at the time of sale, making the logistical factor crucial in shaping the final sale price.

In the various bibliographies consulted, road transport has advantages and disadvantages. Ballou (2006) mentions that the inherent advantages of road transport over other modes are the door-to-door service, with no need for loading or unloading between origin and destination, and no need for transhipment, which is unavoidable in the air and rail modes; it has great frequency and high availability of services, and the speed characteristic of door-to-door service.

As can be seen in Table 1 below, a survey of the advantages and disadvantages of the road mode.

Table 1 - Advantages and disadvantages of the Road Modal

Advantages	Disadvantages
Wide flexibility	Load limited to vehicle size
Used with any type of load	High operating costs
Faster transportation	High risk of theft and accidents
Amply available	Longer delivery times (traffic jams)
Easily adaptable to other modes	High transportation value
Easy to hire and manage	

Source: Logistica para todos (2011).

Looking at the current market (2016), transport companies have to pay high insurance premiums for their cargo, pay high maintenance costs for the equipment they use for transportation, live daily with the loss of materials on the roads and the imminent risk of robberies and accidents due to the infrastructure found on Brazilian highways.

Reis (2006) points out in his studies that the operating costs of a transport vehicle can double in relation to the use of a road in excellent condition to another in a poor state of repair. According to CNT/2010, if all Brazilian highways were properly paved and considered to be in excellent condition, the operating cost of transporting cargo in Brazil could be reduced by up to 25% at the end of operations.

Table 2 shows the total number of paved roads in the BRIC countries. Graph 3 shows the disparity in investment in road infrastructure in Brazil.

Table 2: Total length of paved and unpaved roads in BRIC

Position	Parents	Total length of roads (Km)	Total length of paved roads (Km)	Date of Information
1	≡ndia	4.699.024	2.286.500	2015
2	China	4.460.000	3.746.400	2016
3	* Brazil	1.720.607	213.299	2015
4	Russia	1.396.000	1.005.120	2014

Source: CIA Factbook/2016, CNT/2015. Prepared by the authors.

Graph 3: Comparison of total and paved roads in the BRIC.

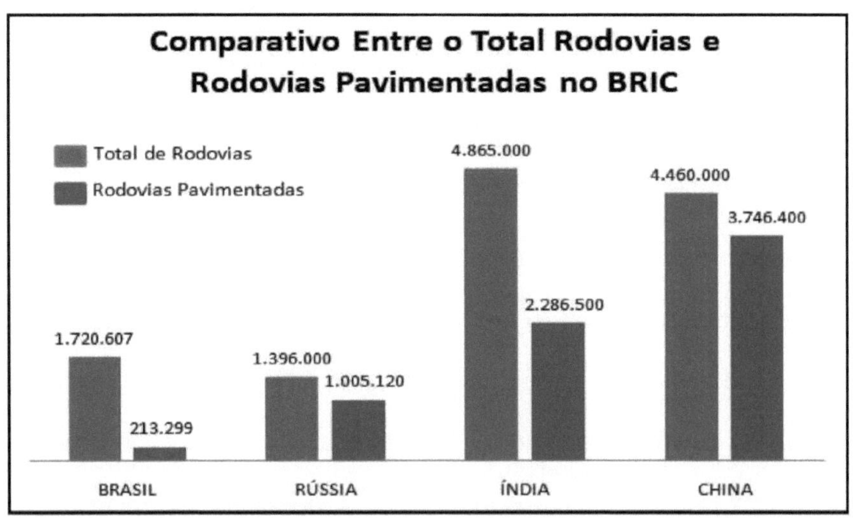

Source: CIA Factbook/2016, CNT/2015. Prepared by the authors.

For a simple comparison of the use of road transport in Brazil, the amount of paved roads in Brazil according to the CIA/2013 is little more than the 191,270 km. of railroads used in China, which shows Brazil's total dependence on road transport and its poor infrastructure.

According to the CIA/2013, road transport in India is aided by 68,525 km of railroads that cross the country, providing support for the better flow of products. Russia has 87,157 km of railroads and, due to its geography, they significantly help the road modal in internal logistics processes.

According to the Ministry of Transport/2015, Brazil currently has 28,000 km of railroad network, not much for a country with more than 8 million km^2 and a road network that is largely in deficit.

3.5 Most transported products within BRIC damage to roads.

Trade within the BRIC is active and accounts for a large proportion of exports and imports among those involved, resulting in frequent and large-scale yogic movements.

According to MRE/2015 (Ministry of Foreign Affairs), Brazil exports basic products to the BRICS, without any type of manufacturing. Unlike Brazil's imports into the BRICS, the vast majority of which import products ready for sale.

According to the MRE/2015, Brazil's main economic partner within the BRICS is China, which accounts for 83% of Brazilian exports, as shown in graph 4. On the other hand, Brazil imports 81% of products from China, as shown in graph 05 below.

Graph 04: Brazilian Exports among BRICS/2015

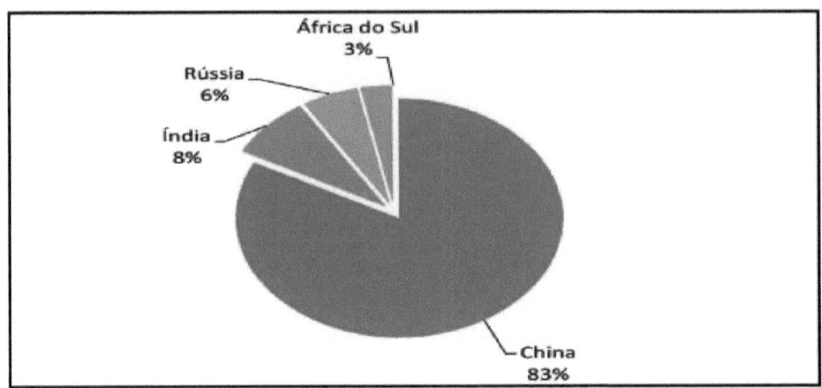

Source: MRE/2015.

Graph 05: Brazilian imports between BRICS.

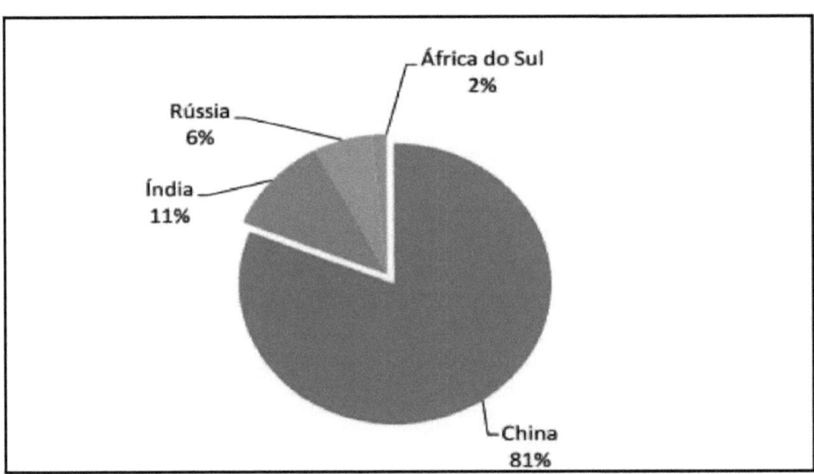

Source: MRE/2015

According to the MRE/2015, of the 83% of Brazilian exports to China, 44.3% refer to soya beans, 19.3% to mineral exports and 11.6% to fuel exports.

MT/2010 presented a report showing which cargoes transported in Brazil have the highest incidence of overweight cases, as shown in graph 6 below.

Graph 6: Loads with the highest number of axle overweight fines

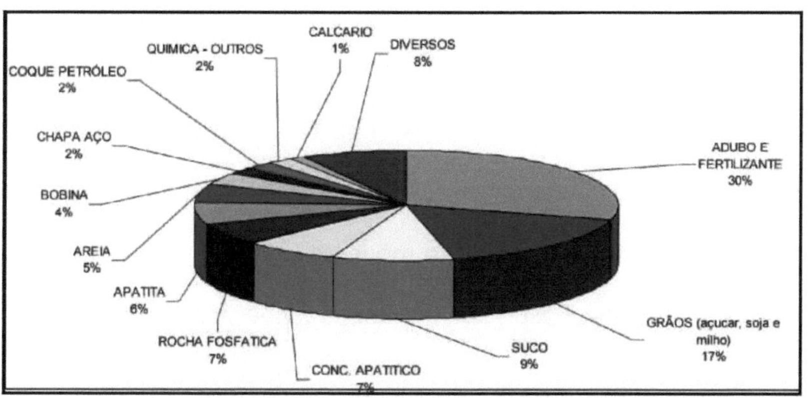

Source: ATR Brasil/2009; ME 2010.

It is clear that in the transport of grains, fertilizers and minerals by road, there is a culture of exceeding the weight of loads in order to be more profitable, and with the increase in production, the incidence of cases tends to increase on Brazilian roads.

Silva, J. P (2007) points out that the transportation of minerals by road results in heavy traffic, causing a series of inconveniences, such as excessive dust, high noise emissions and frequent deterioration of the road system.

High-risk cargo such as fuels and chemicals are constantly on Brazilian roads, exposed to the risks that come with transporting them, as well as cargo that is vulnerable to accidents caused by third parties, transport conditions that are often inadequate and exposed to the open environment, where the risk of accidents increases considerably and increases their proportion, which can trigger a series of environmental impacts, on road infrastructure and often tragedies involving people close to the events.

Brazil has an extensive road network, with around 12% of all roads paved and in usable condition, according to the CNT (2015). Combining this data with little investment in alternative modes, the road mode as the basis for transporting products becomes the only option for farmers, ranchers and industries to transport their produce, and with its excessive use, the wear and tear on the roads and the damage caused to them are inestimable.

Added to this, poor quality asphalt, lack of maintenance and upkeep of roads, construction errors and the frequent overweight of freight vehicles, directly affect the quality of transport routes.

Studies show that just 1% of axle load above the pre-established limit increases sidewalk wear by around 4.32%. Therefore, if the overload is 5%, a road that was designed to last ten years ends up having its useful life reduced to 8.1 years. If the weight exceeds 20%, this durability drops to just 4.5 years (REIS, 2011).

Countries that have a good logistics infrastructure combined with the use of the multimodal concept, where the capacity of each type of transport can be better used, reducing costs and time in operations, increasing competitiveness to meet such demands, will be able to stand out in the world market.

CONCLUSION

By making this economic, geographical and structural comparison of the BRIC countries, we can see that they are on an equal footing economically, which makes the conclusions about Brazil's infrastructure more faithful to the problems encountered here.

The BRIC countries mentioned in this article have complex problems to solve, as they are continental in size and the logistics of transporting materials is a constant challenge for these countries. In addition to this, Brazil makes a serious mistake when it comes to internal freight transportation, relying basically on road transport as the only alternative for transporting its products within the national territory. The rest of the world, on the other hand, uses road transport as an important means of transportation, but it is largely aided by railways that help transport large proportions of its production.

The quality of Brazil's infrastructure, the main focus of this work, is questioned by the quality of public federal highways, which are often abandoned by the government and lack constant maintenance in order to offer a minimum of safety to the professionals who use them.

As discussed in the paper, 12% of Brazil's road network is properly paved, about 213,000 kilometers, and according to CNT/2015, in good condition, and because it is a large country and it is therefore difficult to maintain the quality of these roads, the government opens concessions for extensions of these roads for private companies to manage stretches and through toll collection, keep the road in full traffic conditions. The vast majority of these stretches are in the central and southern regions of the country, where the quality of the federal highways stands out in relation to other Brazilian regions.

The quality of these highways is deteriorating due to the irresponsibility of truck drivers and companies that exceed the weight of their loads and speeds, thus degrading the useful life cycle of the asphalt used on Brazilian roads.

One way out of the problems encountered in Brazil is to increasingly improve the multimodal concept by investing in alternative modes of transport and reducing the movement of large loads by road, which would result in more efficient logistics processes, achieving a significant reduction in costs with regard to vehicle maintenance, reducing the risks caused by long journeys on roads in critical states of repair and a reduction in the value of the goods transported.

REFERENCES

BALLOU, Ronald H. **Supply Chain Management: Planning, Organization and Business Logistics.** Bookman, 2005.

CASELLA, P. B.; BRIC - Brazil, Russia, India, China and South Africa: A Perspective. **Revista da Faculdade de Direito da Universidade de Sao Paulo.** v. 105. January, 2010.

CNT. **CNT Highway Survey 2010: Management Report.** Brasilia, 2010. Available at www.pesquisarodovias.cnt.org.br/. Accessed on 19/09/2016.

CNT. **CNT Highway Survey 2015.** Available at: www.pesquisarodoviascms.cnt.org.br. Accessed on 16/09/2016.

CIA FACTBOOK (2016) **Roadways (Information on Chinese Roadways)**. Available at: www.cia.gov/libraryl. Accessed on 13/09/2016.

CIA FACTBOOK (2016) **Roadways (Information on Russian Highways**). Available at: www.cia.gov/library. Accessed on 13/09/2016.

CIA FACTBOOK (2016) **Indian Roadways**. Available at: www.cia.gov/library. Accessed on 13/09/2016.

FERREIRA, M. A. **Tipos de Modais.** Available at:<http://www.ietec. com.br /site/techoje/categoria/detalhe_artigo/670>. Accessed on: December 20, 2011.

MINISTRY OF FOREIGN AFFAIRS. **Foreign trade between BRICS**. Available at brics.itamaraty.gov.br. Accessed on 14/09/2016.

MINISTRY OF TRANSPORT. **Report on field data research containing technical recommendations complementary to the statistical methodology used in measuring the weight of cargo vehicles.** Available at www.agricultura.gov.br. Accessed on 21/09/2016.

REIS, N. G. (2006). **The Impact of Road Condition on Truck Operating Costs.** NCT &Logistica. Available at www.portalntc.org.br/. Accessed on 12/09/2016.

REIS, N. G. **Excess weight gains manual**. Available at: www.portalntc.org.br Accessed on 18/09/2016.

RIBEIRO, P. C. C. FERREIRA, K. A.; Logistica e Transportes: **Uma discutâo sobre os modais de transporte e o panorama brasileiro.** ENEGEP. Curitiba, 2002.

SANTOS, L. B. BRIC's Role in the World Economy. **Mercator Magazine.** vol. 9, n°19. Fortaleza, 2010.

SILVA, J. P. S.; Environmental impacts caused by mining. **Revista Espaço da Sophia.** n° 08, November, 2007.

TIMM, LEO. **Chinese highways are routes of corruption**. Available at: www.epochtimes.com.br. Accessed on 14/09/2016.

Printed by Books on Demand GmbH, Norderstedt / Germany